RHYTHMS OF THE ECOSYSTEM

JANETTE SHETTER

Illustrated by William Z. Shetter

Pendle Hill Pamphlet 208

About the Author / Janette Knott Shetter was born into a family with an appreciation of nature. She remembers early morning hikes in the foothills of the California coast range as part of her grandfather's routine, while childhood summers spent camping with her parents in the Sierras were livened by her mother's knowledge of wild flowers. Weekends often found the family on the shore of the Pacific Ocean, in a cabin built by her father. Janette's own bent was less for the specifics of botanical names than for the process of growth and the commonality of life. In pursuit of this interest she acquired a doctorate in biochemistry from the University of California, with a special emphasis on bacteria. She has also taken courses in psychology, sociology, and counseling at Indiana University, where her husband, Bill, is on the faculty.

This pamphlet grew out of a course on ecology which Janette gave at Pendle Hill in 1971, when she and her family—husband and two children—were residents and students here. The use of the Dancing Shiva as a central focus owes something to Eugenia Friedman's practice of "holding the poem before the class," and something to Dorothea Blom's insistence on letting the artist's image speak.

ISBN 0-87574-208-4

Library of Congress catalog card number 76-26392

Printed in the United States of America by
Sowers Printing Company, Lebanon, Pennsylvania

October 1976: 3,000

THE DANCE OF SHIVA

Somewhere within the soul of that great god a faint strumming of uncertain sound began to stir. So faintly it began that he turned his whole mind inward and distinguished a slow and stately rhythm rising from his overflowing grace. A hand which held a drum began to tap it in the slow and even cadence of a pulsing heart. His supple body responded to the invitation, and the movement of his legs picked up the rhythm. . . .

As Shiva moved with stately grace and gay abandon, as only a Shiva could do, mist began to form in the outer reaches of the universe. The torch he waved aloft flared forth in pulsing rhythms of the dance, and stars expired before its dazzling brilliance. All that was not Shiva dancing began to fall apart, disintegrate, evaporate into the thin vapors of apparent nothingness. The dance of joy became Tandava, *the Dance of Eternity, a dance of universal death and joy. . . .*

At the climax of the nothingness, when only the audience of gods remained to contemplate his grace and power, Shiva paused, and then began again—as slowly as before. His was to be a dance of the full measure of grace. He had danced the worlds out, now he danced them in again, flinging stars into their heavens, evoking life upon the earth, a kinesthesis of overflowing grace and love. So Shiva danced a new world and a new age into being, and had himself a day of lovely sport and playful joy.

James A. Kirk, STORIES OF THE HINDUS, (New York, 1972), pp. 243-44

THE MYTH OF DANCING SHIVA, god of creation and destruction, has spoken to many generations within Hindu culture, because it embodies a basic truth about human nature and the universe man lives in. Like any great symbol, it speaks to all men and yields up to them such treasure as they find in their own hearts. This pamphlet is the result of what Shiva said to me as a person who had already been made acutely aware of the ecological realities of our time.

In one of his right hands Shiva holds the drum of creation: from his other flows protection. A left hand holds the fire which destroys the cosmos, while the other left points to the upraised foot which symbolizes release, or in Western terms, salvation. There is a man's earring in his right ear, and a woman's in the left. Thus Shiva has a dual nature, and in him there is a reconciliation of opposites.

Likewise within man and his universe there exists a rhythmic tension between the forces for destruction, or reduction to smaller units; and the forces for creation, or synthesis. Let us examine these forces on various levels of complexity: the atom, the molecule, the cell, the individual human being, the groups he belongs to, and the world ecosystem. The Hindus say that the reason for the dance of Shiva lies within his own nature. So it is that the forces for creation and destruction lie in the nature of the universe. What makes this so, and what determines if the dance results in creation or disintegration?

The Dance of the Atoms

In the area of atoms, and the smaller units that compose them, we are in a never-never land where the boundary between matter and energy, which had always seemed so firm, begins to blur. Because of this, the forces that bind matter together into atoms are probably the most powerful ones in the universe.

The atomic reactions most important to earth are those that have been occurring for eons within the sun, but mankind did not become aware of such reactions until the first atomic bomb was exploded, and fission of plutonium into smaller atoms produced huge amounts of destructive energy. Yet not all atomic action is destructive. For in the sun the coming together of two hydrogen atoms at intense temperatures and pressures to form an atom of helium results in the release of an energy which is essential to our planet. This energy has warmed the earth, set the stage for creation, and provided the supporting energy for all life on earth. The dance of the ever-moving atomic particles in the center of the sun could well be described as "ascending leaps of unpremeditated joy" of the Dancing God. This seems like pure creation, for the radioactivity thus produced is safely separated from us by 93 million miles. But in the stock piles of hydrogen bombs on earth is the possibility of unthinkable destruction. For their use would release byproducts whose radioactivity would continue for a longer time than man has existed on this earth. And we could not be sure that the god Shiva would dance the life back on an earth contaminated with deadly, DNA-destroying radioactivity.

In a process where such boundless amounts of energy are released, either for good or evil, there seem at first sight to be no restraining forces. However, we are learning more and more about such matters, and knowledge is a step toward control. The laws of physics, and in particular Einstein's equation, $E=mc^2$, tell us precisely how much energy will be released. And computers can tell us under exactly what circumstances the processes of fission or fusion will take place, for these processes are regulated by properties inherent in the nature of matter itself. As we go from small to larger units in the different levels of organization, we will see that the factors that regulate the dance of destruction or creation become more complex and more difficult to predict.

Cycles of Matter

The continuous, dynamic process in which large macromolecules are broken down into inorganic compounds, which in their turn combine to form fantastically complex molecules is referred to as "cycles of matter." This continuous interaction between the creation of the exquisitely complex and its subsequent breakdown is embodied in the image of the four-armed dancing god. Each dance motion is a creation of something new, never seen in exactly that way before, and each flows into a new form with the destruction of the previous one.

Let us first look at the disintegrative aspect of the dance. Living creatures die, and microorganisms reduce the complex molecules representing the uniqueness of that particular creature into small molecules of nitrate, phosphate, sulfate and carbon dioxide. The universe of that individual creature disintegrates. Yet it is also a dance of life, both for the microorganisms that multiply as a result of decomposition and for plants that could not live if precisely these inorganic compounds were not constantly being returned to the soil and air. Similar decomposition occurs in our stomachs when the complex carbohydrate and protein molecules of the bread we eat are broken down into smaller molecules that can travel via the blood stream to where they are needed.

The decomposition of dead creatures is a foul-smelling process that returns the elements of their bodies to the soil. The Hindus, being aware of this, sometimes describe Shiva as dancing on corpses in drunken ecstasy to the sound of drums made of human skulls, drinking blood from cranium bowls, and gnawing the flesh from human limbs. This image with its inclusion of drinking and eating reminds us that energy is wrenched from the large molecules, and used to carry on the life processes of decomposing bacteria, of other animals, or, if the large molecules are food, of humans. The sulfur and nitrogen from the complex compounds, under the influence of decomposing bacteria, becomes hydrogen sulfide and ammonia. These simple molecules are converted by specialized bacteria into sulfate and nitrate. This is not an act of altruism on the part of the bacteria, but rather a means to their own preservation and growth. Out of the destruction of one form comes the energy for the growth of another, and the new form can be any one of the many in the continuum from bacteria to human.

Intrinsic to the creation of new forms is the synthesis of the most magnificently complex molecules. So complex are the nucleic acid molecules that carry the genetic information, that if they were uncoiled from their precisely organized location within the cell and stretched out, they would be over a yard long. The sound of Shiva's drum, like the *logos* of John the Evangelist, calls forth matter. I find this mythic image of the sound waves creating solid matter a marvelous metaphor for the conversion of the invisible gases, carbon dioxide and water vapor in the air, into very solid plant structures by photosynthesis. Note the magnitude of the process of going from the simple to the complex. The plant, working with 10 to 15 elements in combinations (molecules) of not more than five or six atoms, constructs at least 1000 different kinds of molecules with many of them containing 20,000 atoms. The delicacy and extent of the ordering process in the creation of a living plant from inorganic soil and air molecules is amazing. Further indication of the extent of ordering on the molecular level is the fact that each DNA molecule and many of the protein molecules are uniquely characteristic of the species, and in some cases, of the individual plant or animal that produced it. The rhythmic dance of Shiva with its harmonious, orderly motions, underlined by the beat of the drum, seems a particularly fitting image for form and order. The variety of the forms in the macromolecular world is symbolized by the variety of motions and positions of the dancing four-armed god. With every synthesis of a protein molecule from nitrate and carbon dioxide by plants or from amino acids by animals, disorder is converted to order, and simple structures become complex. With each step of the dance some new, but transient, structure comes into being. It is said that Shiva's gestures are all beings and all words.

And so the cycles go on. Plants use carbon dioxide and release oxygen which is breathed by animals that return carbon dioxide to the air. And exactly the same amount of oxygen is given out in the formation of plant tissue as is required for its ultimate return to carbon dioxide and water, whether that occurs as the result of animal metabolism or fire.

The inseparable aspects of creation and destruction are well known to the ecologist. However, the regulatory mechanism of destruction and synthesis is incompletely known. The rate of destruction, which is also the rate at which nutrients are made available for new growth, determines, in part, the rate and amount of growth that can occur in an ecosystem. Man interferes with the natural regulatory processes, often without knowing it, in many ways. A familiar example is that of the pollution of streams with phosphates from laundry detergents. Under natural conditions, the rate of addition of phosphates to the water via the activities of decomposing bacteria, determines the extent of algal growth. Man's addition of phosphate allows inordinate amounts of algae to grow. Shiva's dance of destruction may be frenzied, but it is still a dance with a beat—a regulated rhythm, that balances the extent of the concomitant creation.

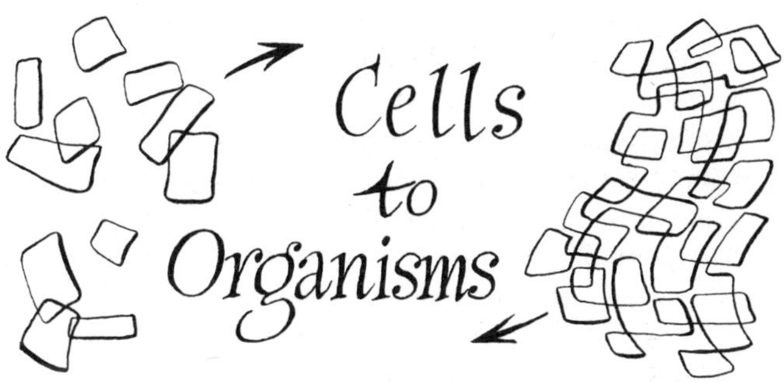

Cells to Organisms

Somehow, millions of years ago, an aggregate of cells developed a survival power not available to unicellular creatures, and the slow course of evolution took a leap forward with the advent of the multicellular way of life. Let us first look at the disintegrative possibilities within multicellularity. The forces holding cells together in an integrated whole can be tenuous. With the beat of Shiva's drum renegade cells within a multicellular creature can return to the unicellular way of uncontrolled reproduction, and there may be a cancer. A cancer may not appear to be unicellular, but it is so in the sense that this collection of cells no longer is an integrated part of the larger organism, but rather an unregulated, each-cell-for-himself mass.

The forces for unicellularity are not always destructive; they can also be constructive. This is most obviously true in our own bodies where single cells are produced that have the potential of uniting with another single cell from another human being, and a unique new life begins in the womb.

Having acknowledged the trend in the direction of single-celled organisms, let us now consider some of the factors in the opposite drive towards complexity. All living beings need some means of acquiring nutrients, eliminating waste products, protection, reproduction, and awareness of their environment. Multicellularity allows different cells to specialize in performing these different functions. These specialized cells can be much more effective in their different activities than a single cell that has to meet all these needs. Thus a multicellular organism is better able to survive than its single-celled ancestors. The survival value of specialization is a powerful force in the direction of the multicellular and the more complex. Slime molds represent an experiment with multicellularity. Here the forces for specialization function only at the time of reproduction when single cells aggregate, piling on top of each other, lifting the reproductive structure upwards so that the spores, with the aid of the wind, can be disseminated into new territory, where they do not compete for food with the mother cells.

At the other end of complexity from the slime mold is the human being made up of billions of cells with specialized functions with each cell contributing to, and playing an integrated part in, the whole. Indeed, the growth of a single fertilized cell into a fully grown human being is of miraculous complexity.

It would seem that larger and more complex multicellular creatures have all the advantages. The variety and success of the multicellular way suggests that there is a powerful evolutionary force in this direction. How is it that single-celled bacteria, algae, protozoa and yeasts have survived? Single cells have a less complicated reproductive system with the result that they can form two cells in a relatively short time—for some bacteria, every twenty minutes. Whenever nutrients become available, for example, when an animal dies, the bacterial population can reach huge numbers in a short time, and the dead body becomes inorganic nutrients for plant growth with the operation of the cycles of matter. Also, single-celled organisms can live in extreme temperatures from near freezing to about 180°F, and can derive energy from unusual chemicals such as sulfur and cellulose. It seems as if a cell must give up biochemical complexity to achieve the structural complexity of the multicellular.

Just as the question of whether the issue will be synthesis or destruction is a pressing one at the molecular level, so is it on the level of cellularity. How a single fertilized cell multiplies to form specialized organs is an active field of biological research. How does a cell "know" that it is to become a nerve cell? Where does it get the set of plans for being nerve-cell shaped? Although some aspects of the answers are known, this is still one of embryology's unsolved mysteries. Shiva's drum beats, the contractile vacuole of the protozoan pulses, the heart of the vertebrate pumps, the plants respond to the seasons, each cell and organism according to its own rhythm; and when the rhythm is violated—that is, when the regulation goes awry—a cell goes wild and a malignant growth begins. A pollutant released into the ecosystem, such as the defoliating agent the

United States used in Viet Nam, can result in increased incidence of uncontrolled cancerous growth. Man is busy making and using new compounds never before present on earth and releasing them into the air, water, and soil. Life evolved over millions of years in response to the chemicals naturally found on earth. It is entirely possible that any new compound, given long exposure, is detrimental to biochemical regulatory systems, and many are known to be.

But the cause of cancer is complex. Certainly heredity plays a role, and so do viruses—probably by interfering with the regulatory DNA molecules. There is also some evidence that cancer is more likely to occur in persons who are afraid to express hostility and anger outwardly and so turn it inward upon their cells. All these in some way appear to play a role in the beat of Shiva's drum in the direction of the return to the unicellular way of cancer. The power of the evolutionary force toward the multicellular is manifest, but again, we have much to learn about its nature.

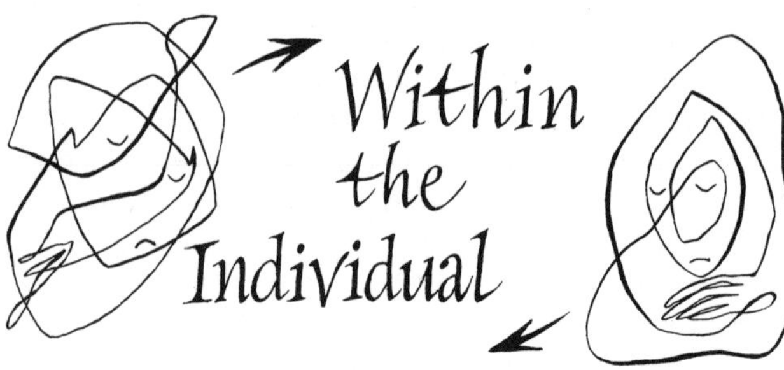

Within the Individual

Here we leave the area of measurable forces. In speaking of atoms, macromolecules, and multicellular growth, we were in an area where the results could be calculated. But with the forces within ourselves an exact determination is not possible. However, we can recognize the extremes. A person in a mental institution without a clear concept of who or where he is represents an extreme of disintegration. Wholeness is represented by the creative, mature person, actively aware of and concerned about where he is.

There are two types of forces that produce psychic disintegration. The first is the more easily understood because it represents chemicals acting, or not acting, on the human nervous system. Lead and mercury, increasingly common pollutants, act by poisoning nerve cells. Or a lack of certain compounds may result in malnutrition, where the central nervous system does not form properly because of insufficient protein in a child's diet.

The second type of force is the more subtle effect of social environment, as when the struggle for survival is so intense that there is barely enough energy to stay alive, leaving none for the drive for wholeness. Or the social nexus in which a child is placed may have an extremely detrimental, disintegrating effect if it is one of continuous contradiction, where for example a parent's words say one thing is expected from the child, while his actions insist upon another. And to deny a child's attempts to reach out to people, to be creative, and to express his being also represents a force tending away from integration.

But, as we have seen, the forces of disintegration are not always negative. Shiva had to dance the universe into seeming nothingness before he could create. A clinical psychologist tells me there is a healing psychosis in which the patient must get worse before he can get better. I have been privileged to see the "inflexible" ideas of a retired person dissolve and change from under the loving pressure of those younger than himself. Unless we welcome Shiva's painful dance of destruction, there is no chance for creative flexibility. This is part of the meaning of "he who would save his life will lose it, and he who would lose his life will save it.'' Is not ''losing one's life'' a breaking up of the old ways so that a new more unified whole can be formed? I suspect that any religious experience involves the breaking the bonds of the old, dissolving the boundaries between man and God.

Openness, flexibility, willingness to go in new directions; all of these require that the god Shiva painfully use our soul as his dancing floor as he dances old prejudices out, and disintegrates our old preconceptions, breaking them down to provide the building blocks needed for the new synthesis.

Integration, creation, synthesis are all positive words. Yet they can have their negative side, when they stunt further growth. The thrusting of responsibility onto a child at too early an age can do this. When a prescribed definition of wholeness is assumed and creative deviance is prevented, the integration resulting is unhealthy. Associated with this is the sort of ''integration'' that insists on consistency.

Healthy aspects of the integrative process or the synthesis of the intrapsychic forces working within us have been emphasized many times in this pamphlet series, as when Dorothea Blom says, ''The side of ourselves that we reject goes underground and becomes subversive, a threat to ourselves and our world.'' Mary Morrison speaks of ''the self calling to the self.'' There are pairs of seeming opposites in all of us; the masculine and the feminine, the dependent and independent, the contemplative and the active. All of these cry out for an integration, a bringing together and giving honor to each member of the pair.

Little is known about the regulatory forces here. My belief is that there is a force in the direction of psychic wholeness just as powerful as the evolutionary force toward multicellularity. A. Maslow and some others engaged in the psychiatric profession see their role as releasing this force. To describe it Friends might use the term ''God Within.''

A regulatory force on the side of disintegration, analogous to the ecological need for the decomposing bacteria, is society's need for its deviants. These can serve to strengthen societal norms, as examples of what is *not* done. And occasionally from a deviant, as from a mutant, comes a creation that points in a new direction.

How do we know when it is "a time to build" and when "a time to tear down?" Shiva "turned his whole mind inward and distinguished a slow and stately rhythm rising from his overflowing grace." We too can turn our minds inward and attempt to be sensitive to our own inner rhythm. Or, to put it in the familiar Quaker terms, be mindful of the Light Within.

Individuals and Groups

Shiva's mystic dance of dissolution and creation is repeated every day in the patterns of our lives. We sleep and dream alone, a solitary individual, not responding to anyone who may be in the same room; we awake and find ourselves a member of a family or some other unit. And so the pendulum swings between drawing apart as an individual, and becoming a part of a group; the family, the Meeting, a work party, a committee, or people talking over coffee. On this level of organization there is more room for human choice to enter in. And if we listen sensitively to the beat of Shiva's drum, we will find that as individuals we each need differing amounts of solitude and of time when we are a needed member of a larger unit.

Within every group there are the forces for both cohesion and disintegration, and both have their positive and negative aspects. First let us look at those elements acting to decrease group cohesion which generally have negative results. There is the dominator, the person who knows what the group should do and who insists that everyone do it, and who does not listen to the natural rhythm of the group. Then there is the force, often contagious, unwittingly exuded by an uninvolved member who has just come to watch, but who doesn't intend to do any of the work or to really let the others know who he is. And there is the member whose words and actions keep saying "pay attention to me." Sometimes a group can be helpful to such a person if he or she is able to admit to a deep hurt, and if the group has patience, love, and concern. Occasionally forces from outside act in the direction of disintegration, as when a committee is given an unclear or impossible task, or when members have schedules which make a meeting time impossible. Finally, there are forces, generated not by a single member but by the composition of the group, that act towards its dissolution, such as the uneven preparation of members for the task at hand.

But forces of dissolution can be positive, as when an individual withdraws from a group so that he can become more aware of who he is as an individual. Some rebellion of teen-agers, hurtful as it can be to parents, represents the growth force of an individual separating from the family as he becomes more fully mature. The Friend who does not nod "I approve" at every item that comes up at Meeting for Business, and who resists the easy, less time-consuming consensus, can be a very positive force, as he makes the Meeting really think about its decisions. Within each individual lies a drive toward competition, self-assertion, self-expression, control of others and the environment, and individuality. This is the force that has led to the science and technology of the 20th century. It is not a force that leads to group cohesiveness, but when balanced by cohesion-promoting forces, so that it does not become ruthless, exploitative and arrogant, it can be a powerful forward thrust.

Just as there are positive aspects to dissolution, there are negative aspects to group cohesiveness. There are those who hear in the beat of Shiva's drum only an intoxication with the group. These people submerge themselves in a group, losing their own identity, and unfortunately there are groups whose norms encourage this. Here is the conformist who looks to others for his direction, the person who seeks an easy consensus. Research indicates that most people will deny what they see with their own eyes if several people whose judgement they trust say it is not so. The norms, or accepted ways of behaving within a group, have a strong influence upon members who know they will be rejected if the norms are not followed. This negative force is well known to parents of certain teenagers as they see the effects of their children's peer group.

Research studies point to some factors that promote cohesiveness. It has been found that groups that are difficult to join are more highly valued than those that accept anybody. This fact has something to say to those who would make it easier to join the Society of Friends. Groups that have a clearly defined common goal are more cohesive. Groups that succeed in achieving their goals are more highly valued by their members than those that don't. Meetings might remember this when they hesitate to undertake some social action project. In cohesive groups each member has a role, that is each person knows what is expected of him or her, and feels assured of being able to carry out the expectations. In Quaker terms I think this means that each Friend has something unique to contribute to the Meeting. The Meeting, in its turn, lets him know that his contribution is needed, encouraging him to develop less-used potentials and do things he hasn't tried before. This sense that each person has a place, a role, or a function due to unique abilities can bring a releasing sense of belonging and an opportunity for growth.

Sociology books don't talk much about "love." But my experience suggests that love and concern for others is the main cohesive force that holds groups together. When these are present, there is a willingness to bear the discomfort of working through conflict, and keeping the lines of communication open.

When accompanied by love group norms can be very positive. This is the way children learn from their families how humans are expected to behave. In a group where people are listened to and appreciated for the unique individuals they are, things happen. People become free to reveal themselves as they had not previously been able to do. They see potentials within themselves that they didn't realize were there, and become more free to respond to the Light Within. And they find courage to act, using the newly discovered forces within themselves. The joy and some of the pain of this process is expressed in a Hindu poem about the dance of Shiva.

> O, my world, my life, my blossoming, my ecstasy!
> Your every moment I create
> By negation of all forms previously lived through:
> I am eternal negation . . .
> Enjoying this dance, choking in this whirlwind,
> Into the domain of ecstasy, He takes swift flight.
> In this unceasing change, in this flight, aimless, divine
> The Spirit comprehends Himself,
> In the power of will, alone free
> Ever-creating, all-irradiating, all vivifying,
> Divinely playing in the multiplicity of forms, He
> comprehends Himself . . .[1]

---

[1] Ananda K. Coomaraswamy, *The Dance of Shiva* (New York, 1957), p. 75.

In the coming together of smaller units to create more complex structures on the atomic, molecular, and cellular levels, the pattern of the Dance of Shiva can be clearly seen. With individuals, and with the forming of groups by individuals, free choice is a dominant factor. However, the sum of many free choices forms a pattern, so that the factors involved here are no less a part of Shiva's dance. But as every Friend knows who has deeply struggled with how much time to spend alone in prayer and meditation, and how much to contribute to the various groups of which he or she is a part, the only true guide is a sensitive attunement to the leading of the Spirit—Shiva turns his whole mind inward.

Mankind and the Ecosystem

Let us now look at mankind as a whole relating to the planet upon which he lives. There are three main ways in which mankind can consider the earth: 1) As a domain which is his to exploit for the benefit of the present generation. 2) As a natural resource, to be technologically managed for the benefit of future generations. 3) As the creation of the Spirit of God, whereon he lives in harmony with nature, rather than merely using nature.

Unfortunately, to distinguish between "living in harmony

with" and "using" may not always be possible, but I submit that there will be a difference in the long-range effects of the man who designs a hydroelectric power plant with a love for the beauty of the canyon, over that of the man who considers the canyon to be there only so that he can use the water. As I was writing this, Haydn's *Creation* came on the radio, and that oratorio is a magnificent expression of the awe at the wonders of God's creation that is at the root of living in harmony with the earth and its creatures. When mankind does live in harmony with the earth, a new level or organization exists, the world ecosystem. There is, to be sure, a world ecosystem of sorts, no matter how mankind lives, but it is unstable, out of balance, or disorganized when man wittingly or unwittingly plays a disruptive, rather than a harmonious role in it.

Man is only just beginning to become aware of the destructive effects he can have upon the earth. He has watched acres of productive land become permanent desert because of overgrazing. He has seen species become extinct within one man's lifetime. He recognizes that it is possible to pollute the ocean to such an extent that it loses the ability to nurture life. With the life-carrying power of the ocean's three fourths of the earth's surface gone, the rest would follow. With atomic weapons "all that is not Shiva dancing" may indeed, "fall apart, disintegrate, evaporate into the thin vapors." Generally there is a great deal of stability in food chains and the other intricate interrelationships of living beings, but we don't know how much pressure they can take. In short, man does not know what he is doing. Witness the disagreement among scientists on the effect of the increased carbon dioxide in atmosphere produced by burning fossil fuels. If all known fossil fuel reserves were burned, the carbon dioxide concentration in the atmosphere would double. Present life is adapted to the existing concentration. We simply do not know what the effect of doubling the amount of carbon dioxide would be, but it is certain that it would have some effect. An obvious disintegrating force is the uncontrolled growth of the human population, whose course is closely analogous to the growth of a cancer within the body. In both cases many different specialized cell types (species) are crowded or starved out by a homogeneous population of cancer cells (human beings). In both cases where there once was strict regulation of the growth of the cells, or the size of the population, the control has gone astray. And in both cases death of the body containing the cancer or the ecosystem earth is the inevitable result if control is not restored.

These are only a few of the many hazards of the disintegrating forces acting on the once-strong ties between man and the earth he lives on. Let us now look at more positive aspects of Shiva's dance of destruction.

One right hand of Shiva is in a gesture of preservation. To an ecologist preservation represents equal amounts of destruction and creation. When creation proceeds too rapidly, as in the present rate of human population growth, preservation means destruction. By this I mean that if something is not done at some point to cut down on the population the ecological balance may be so badly upset as to have reached a point of no return, because of polluted and impoverished soil, atmosphere, and water. The devotees of Shiva had a deep intuition of the necessity for destruction. Death is a necessity on an earth with a limited amount of life-sustaining elements. If population control cannot be achieved by other means, it may take place by means of the horrendous forces of starvation and plague. The Black Plague killed approximately half of the population in Europe, and overcrowded conditions are the soil that grows pestilence. Would we be able to see the inherent necessity of such suffering?

Hidden deep within the destruction there is always the hope that not just new, but perhaps also more complex, more conscious, and more unified forms might arise. This hope which is expressed by Teilhard de Chardin and others arises more from Western than Indian soil, but it makes it easier for us Westerners to truly welcome the dissolution of that which we hold dear. Those who have seen dissolution in their personal lives prepare the way for new growth and development that could not have otherwise taken place, may find the frenzied dance of destruction somewhat easier to accept. Others might see in it the shattering blow of the redemptive catastrophe.

An overview of the history of the human race gives us a look at another positive aspect of Shiva's dance of destruction. The loosening of primitive man's bonds to mountain and fertility gods brought with it a decrease in fear and superstition, and a freedom to create in new ways. When men released their ties to the land and moved to cities, they created civilization and culture, which could never have happened in the smaller farm-bound units, but at the same time they were moving away from harmony with the land. When man released energy stored for thousands of years in oil, he brought rapid movement and communication, and a decrease in human misery and back-breaking toil. Surely we do not have to turn our backs on all of these advances in order to form a viable man-earth ecosystem, that does not have within itself the seeds for its own destruction.

I suspect that there is an analogy to the relation between the individual human being and his mother, and that between mankind and Mother Earth. At first a baby does not distinguish between itself and the breast that feeds it, and so, I think, it was for primitive man who thought of the earth as part of himself. And just as the child must separate himself from his mother in order to become a man, so it is that mankind has separated himself from the earth, perhaps a bit like a rebellious teen-ager failing in appreciation for what his mother continues to give him. As the independence-loving teen-ager finds that he cannot survive without cooperating with his fellows and that he cannot truly know himself until he comes to know his parents on an adult level where he is a separate person, so it is I suspect with mankind who, aware of his separate unique quality as man, must rediscover and renew his appreciation of the magnificent bounty, terror, and beauty of Nature.

We have seen that there are both positive and negative aspects to all of Shiva's rhythms. Shiva's drum beats for disintegration, and also for reunion with the earth. Can there be negative aspects to forces driving man to retie his bonds to the earth? What has been said in the preceding paragraph suggests that there are. Man simply cannot go back to the primitive close-to-the-earth farming methods. The earth is already too over-populated for the human race to live by those methods. Man cannot go back to believing in earth, mountain, and fertility gods and goddesses; his mind is too sophisticated for that. Man now does control nature in many ways and he cannot give up that control. Having eaten of the Tree of Knowledge, he has been driven out of the garden, and the gate to Eden is guarded by angels with flaming swords.

Let us listen carefully to the dimly heard beat of Shiva's drum and see if we can hear the beat for a new integration of man, separate, and uniquely himself, with his environment, an integration that is sensitive to the inner rhythms of the ecosystem and harmonizes with them. These rhythms are complex interacting living and non-living systems of energy flow, cycles of matter, and water. Understanding them takes the patient genius of scientists, appreciating them the sensitive perceptions of the artist and poet, and disciplining ourselves to listen to them and order our lives in harmony with them may well take the dedication of the saint. Again we look to the dancing god. "Shiva also dances in the hearts of his devotees as the Great Soul. As he dances, one's egoism is consumed and one is rendered pure in soul and without any spiritual blemish."[2]

[2]Oroon Ghosh, *The Dance of Shiva and Other Tales from India* (New York: The New American Library, 1965) p. 109.

Man is set down within an ecosystem; its order is also his order. Not only are integration and disintegration an intrinsic dynamic of the ecosystem, they are also acting within each of us. Seeing ourselves as an aspect of a universal order has a dissolving effect on our egoism. When one's egoism begins to be consumed, one becomes less strongly bound to what from the viewpoint of the ego is good or bad. Then one is enabled to watch the dance of personal integration and disintegration with joy as well as fear.

"Shiva dancing in the hearts of his devotees" means becoming aware of these forces within ourselves, and learning to accept and cooperate with those painful processes that disintegrate our self-centeredness. Being able to recognize the need for, and to welcome, all of Shiva's forces leads to the freedom that comes from accepting what is.

> The dancing foot, the sound of the tinkling bells,
> The songs that are sung and the varying steps,
> The form assumed by our Dancing Shiva—
> Find out these within yourself, then shall your fetters
> fall away.[3]

---

[3]Coomaraswamy, p. 72.